RAISING MONARCHS

Caring for One of God's Graceful Creatures

RAISING MONARCHS

For information about special discounts for bulk purchases, please contact Sunbury Press, Inc. Wholesale Dept. at (855) 338-8359 or orders@sunburypress.com.

To request one of our authors for speaking engagements or book signings, please contact Sunbury Press, Inc. Publicity Dept. at publicity@sunburypress.com.

FIRST SUNBURY PRESS EDITION
Printed in the United States of America
June 2013

ISBN 978-1-62006-246-3

Published by:
Sunbury Press, Inc.
50 West Main Street
Mechanicsburg, PA   17055

www.sunburypress.com

Mechanicsburg, Pennsylvania   USA

For my Mom and Dad and Mike and Michael

Touch it before its wings are freed
And leaves us lost in mystery.

I look out the kitchen window. Suddenly I see a flash of orange. Oh my God, she is here! She circles around, and then descends. I race to the back door and speed through the porch. I hide behind the big blue spruce. My heart pounds.

I slowly step around the tree. She's right there! I approach and kneel beside her. She remains. I watch as she bends her lovely black abdomen. We are silent, but I swear the universe sings as she deposits her tiny white egg on the underside of a green milkweed leaf.

Each spring I wait for the first female monarch butterfly to appear in my suburban Chicago backyard to start a new cycle of life. And it all begins with the emergence of milkweed. The appearance of milkweed in my garden is just as thrilling as the arrival of the first female monarch.

The newly sprouted common milkweed (pictured on right) attracted this early-spring female. I watched her deliver an egg before flying away.

As the warm winds start to blow across the Midwest in April, I run my fingers through the dirt and scan the garden soil for evidence of milkweed.

The female monarch usually deposits a single egg on a single milkweed plant.

I brought milkweed into my yard more than a decade ago. My son, Michael, was young then and he and I planted our butterfly garden together. Milkweed, known as the monarchs' host plant, is the only source of food that monarch caterpillars will eat. Without its availability in my yard and across North America, the monarch butterfly life cycle would cease.

Once we planted the milkweed, Michael and I sat back; we thought our work was done. We dreamed about our garden coming to life. We hoped to see a habitat crawling with yellow, black, and white-striped caterpillars.

Miraculously, the habitat did attract elegant egg-laying female butterflies, but locating the colorful caterpillars was a different story. We wondered why, so we turned to our library. We checked out bags of books and studied monarch life. We read that less than ten percent of monarch eggs and caterpillars survive. Eggs, the first phase of the monarch's four-part life cycle, and caterpillars, the second part, are at the mercy of many predators. Spiders suck out the contents of eggs and various other insects feast on or infect the tiny caterpillars.

After we learned this, we decided to start our rescue operation. We turned over leaves and looked for eggs. We noticed many eggs already had been emptied by spiders, but we collected the leaves that held viable eggs.

We placed the leaves over the side of a pie tin so a part of the leaf could reach a small amount of water in the bottom of the tin. That way the egg-bearing leaves could stay fresh. We kept our pie tins on our screened-in back porch so the temperature matched the outdoors.

Monarch eggs on leaves stay fresh in a little bit of water.

Michael and I had learned that females can lay hundreds of eggs, which are white, elliptical, and about the size of a sprinkle on a birthday cupcake. Books informed us that the eggs hatch in a few days depending on the temperature, that is, the warmer the weather, the quicker the caterpillar develops. With that information, we watched our eggs. And after a couple of days, we noticed a change. We observed that the tip of each egg had darkened. We realized that each dark tip was a caterpillar's head, which we were soon to see!

Our newborn caterpillars were tiny; they were about the length of a hyphen. With the help of a magnifying glass, we watched the caterpillars, also called larvae, eat their own egg cases and then start to devour the milkweed leaf. We noticed that a baby caterpillar often chewed the shape of the letter "C" in the leaf as it took its first bites.

This young caterpillar seems to prefer the letter "O."

Little by little, we came up with a caterpillar-rearing system. Once a caterpillar would hatch, we transferred it to a milkweed cutting, which was kept alive in water (on left). Then we would place the cutting and the caterpillar in a large plastic container outfitted with a screen at the top and on the side for sufficient air circulation.

This container is complete. It holds a milkweed plant and a small caterpillar. One caterpillar can eat five to seven large milkweed leaves. The plant gets its name because its leaves are filled with milky sap.

Caterpillars eat for about two weeks.

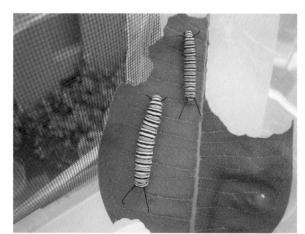

We named our first caterpillars "Fudgy" and "Budgy" and have recycled those names ever since.

Caterpillars grow so much that they shed their skin five times. They eat their discarded skin because it provides nutrients.

A full-grown caterpillar crawls across Michael's hand. They can measure up to two inches when mature.

A pair of friendly caterpillars share the same leaf.

Striping patterns change a little every time a caterpillar sheds its skin.

Each summer, Michael and I witness not only how much a monarch caterpillar eats but how much frass it generates. Frass is caterpillar poop! We clean out our containers several times a week. If any of the caterpillars need more food, we visit the milkweed garden for a fresh supply.

The common milkweed in the backyard basks in the sun. Its large showy flowers bloom in June. The plants attract egg-laying females from May through September.

The milkweed in May is big enough to sustain dozens of hungry monarch caterpillars.

Milkweed plants contain a toxic chemical called cardiac glycosides, which a monarch caterpillar ingests and stores in its body. This chemical is retained when the caterpillar becomes a butterfly. Both caterpillars and butterflies are protected from some predators because of this toxin.

Once a caterpillar grows to about the size of an adult pinkie finger in about ten to fourteen days, it stops eating. Now, its task is to explore its surroundings and choose a safe spot to become a chrysalis, or pupa, the third part of its life cycle. After it settles in one place, the fat caterpillar will become still for a day and begin to make a silky web and knob. It secretes silky threads from a gland near its mouth. Eventually, it will suspend itself from the knob and hang in the shape of the letter "J."

Many caterpillars that we raise form their chrysalises on the underside of the lid of the container in which they are housed. However, some prefer to attach themselves to the underside of a milkweed leaf inside the container like this one (above). Part of the silky knob is visible at the caterpillar's rear end.

Several caterpillars in the "J" position prepare to become chrysalises. Many hang from the lid of the large plastic container in which they were fed.

The caterpillar will hang in the "J" position for about a day. What happens next is truly amazing. The suspended caterpillar will start to wiggle and its outer layer of striped skin, called its cuticle, will split open at its head. The skin splits and peels right off the caterpillar all the way up its back to where the rear legs are attached to the silky knob. All this happens within a couple of minutes.

The shriveled skin and frass lie below the brand-new pupa.

As the outer layer of skin drops, the green body of the caterpillar is exposed. After an hour or two, the body hardens and becomes one of Mother Nature's most stunning works of art. The monarch chrysalis looks like a piece of jewelry.

The monarch chrysalis is exquisite.

Each chrysalis resembles a precious gem dotted with gold.

This newly formed chrysalis is attached to the milkweed's main vein where it is well-camouflaged.

Two monarch chrysalises attached themselves to the underside of the common milkweed plant.

New chrysalises decorate the lid of a plastic container. The garden in the background will provide enough nectar for the butterflies-to-be.

New chrysalises are the color of jade. As the time of emergence, or eclosure, nears, the colors and wing pattern become obvious.

It is rare to find monarch chrysalises in the garden. I have seen only one chrysalis in my milkweed patch during the past dozen years. Full-grown caterpillars usually leave the habitat and secure themselves to branches or twigs or some other structure that is removed from predators.

The chrysalis stage ranges from ten to fourteen days. A day or so before the monarch butterfly emerges, the green chrysalis usually turns to black. This is because the chrysalis has become transparent and the monarch's black body and wings are now visible. The color change signals the imminent birth of the butterfly.

This chrysalis is just starting to split. The one pictured below is also opening. Most chrysalises are firmly secured to withstand the outdoor elements. The silky web is widespread on the ceiling.

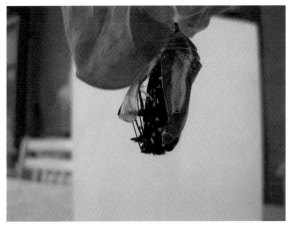

During the mid-summer months, monarchs usually emerge from their chrysalises in early morning. Witnessing the birth of a monarch butterfly is extraordinary. Its legs push open a portal and its swollen abdomen drops and its folded wings are freed. The butterfly enters the world upside down. While hanging in this position, it immediately begins to pump fluid out of its enlarged abdomen into its rumpled wings. Within three or four minutes the wings straighten and the butterfly, although hanging upside down, is picture-perfect.

Butterflies sip nectar from flowers with a straw-like tube called a proboscis, which is visible in the photo above. A minute or two after birth, a butterfly needs to uncurl its proboscis to get both parts working together so it can drink nectar properly. When not drinking, the proboscis is rolled up.

This
newborn
monarch
is less
than five
minutes
old. Its
wings are
still
wrinkled.

Even though these monarchs look ready to fly, their wings need more time to dry. Excess brown fluid continues to drip from their abdomens onto a paper towel.

After several hours, the butterflies become active and we release them.

This butterfly isn't quite ready to fly yet and prefers to remain upside down to drip and dry.

The butterfly, which is the final phase of the four-part monarch life cycle, is also called the adult stage because a newborn butterfly is already full size. New monarchs need to remain suspended for several hours so their wings can unfold properly. After an hour or two, the butterfly will open and close its wings for the first time. Such a spectacle to witness!

Once a butterfly begins flapping its wings and moving around, we prepare for its release. We usually carry the container outside and lift the lid. Oftentimes, the monarch senses freedom and will take off. We'll watch it fly. As it gains altitude, it stops flapping and gracefully glides. It usually soars toward a few tall trees behind our yard and rests there. A monarch can remain up there for several hours. It's a safe place to practice opening and closing its stained-glass wings.

Some monarchs, however, do not want to take flight right away. We have a special place in our yard for them. We set them at eye level on our big blue spruce. Stationed there, they continue to dry and practice flapping their wings.

Butterflies do not need to start drinking nectar right away. They can go for a day without food. We avoid placing newborns on flowers because bees can attack them. The monarch butterflies that are born in the summer live for only two to six weeks.

This female monarch continues to dry and harden before she can take her first flight.

This monarch is a male because it has two black dots on its hind wings near the tip of its abdomen. The dots are scent glands to attract the female. The male's black veins are narrower than those of the female as is seen in these photos.

After the birth of a butterfly, its empty cellophane-like chrysalis remains. When we were cleaning this lid above, we were surprised to discover that the silky threads from more than eight chrysalises were attached to one another as shown below. A chrysalis kite!

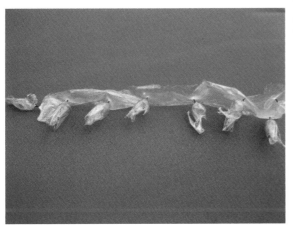

As the summer progresses, the monarch population grows. I have seen four or five different pairs of mating monarchs in my yard during the past few summers, just like this couple. When disturbed, they fly off but remain attached with the male carrying the female.

During the past several summers, I have opened up my backyard to the community and encouraged visitors to walk through the monarch habitat. I teach people about monarchs and give away dozens of caterpillars to families, children, and adults so they can enjoy the intimate experience of raising one of God's most graceful creatures. It's always a bonus when a monarch butterfly suddenly flies through the yard during a presentation. The audience realizes that butterflies are still around even though they may not have seen one in a long while.

After attending a presentation, people say they are grateful for the opportunity to see all four stages of the monarch life cycle. They also start to realize how important native plants are to the insects that live among us. Weeds become valued for the essential role they play in the lives of butterflies.

I prepare containers before the presentations. We cut out the tops and sides of containers and add screening. Friends donate used pretzel jars and screens.

These containers are ready to give away.

I bought some large containers and turned them into caterpillar condos. I can raise up to ten caterpillars in each one.

After getting up close to a monarch, many people decide to add milkweed to their gardens. They, too, want to provide a habitat for these wonderful creatures. I have dug up dozens of milkweed and handed them out to people young and old.

I was thrilled to spot this full-grown monarch caterpillar (center) feasting outside on a swamp milkweed plant in my yard one August. When I do find large caterpillars in the garden, which isn't very often, I leave them there and hope that they can continue to survive on their own. While I could not locate this caterpillar the next day, it may have been because it wandered off to a remote spot to hang and form its chrysalis. At least I hope it did!

Swamp milkweed (to the left of and behind the pink coneflowers) offers gorgeous, scented pink flowers in late summer and does not spread as quickly as common milkweed. Its leaves are much smaller than the common milkweed.

I grow common milkweed because its leaves are big enough to feed the couple hundred monarch caterpillars that we care for each summer. Once established, common milkweed can spread easily, but can be controlled by simply pulling up any unwanted plants.

In addition to common milkweed (*Asclepias syriaca*), I grow swamp milkweed (*Asclepias incarnata*) and butterfly weed (*Asclepias tuberosa*). All three of these varieties of milkweed are beautiful and serve as host plants for the monarchs. More than 100 types of milkweed grow in North America.

This is a monarch egg on the top side of the swamp milkweed leaf.

A monarch dines on the lovely, ball-shaped common milkweed flowers.

When in bloom, common milkweed offers a strong sweet fragrance and attracts monarchs, black swallowtails, tiger swallowtails, mourning cloaks, question marks, red admirals, commas, and hummingbird clearwing moths. The backyard sometimes feels like a butterfly museum!

Before the milkweed flowers bloom, female butterflies often lay their eggs on the tightly packed clusters of flower buds as shown. In early June, I find more eggs on the flower buds than on the underside of milkweed leaves. Once the caterpillar hatches, it crawls in between the cluster of buds and feeds, unseen by predators.

Common milkweed flowers last from June to July and then pods appear.

Milkweed pods mature and split open in September in the Midwest.

Monarch butterflies that are born in late August and September are different from the ones that emerge earlier in the summer. They are born with an immature reproductive system and are unable to mate until much later in their lives. Unlike the summer monarchs, they do not glide around my yard as if they had not a care in the world. These final-generation monarchs face an incredible journey.

Their job is to fatten up so they can fly thousands of miles from their place of birth to the mountains of central Mexico. There they will gather together on the trunks and limbs of towering oyamel fir trees and wait out the winter. These monarchs can live up to eight or nine months, not just a few weeks like their parents and grandparents.

These late-summer or migratory-generation monarchs born in my yard are fixated on one thing: food. My job is to make sure I have plenty of late-blooming flowers for them to dine on.

For two days, this stunning male monarch sipped nectar from the sweet autumn clematis that blooms in my yard in September.

This monarch, along with a handful of others, nectared on my verbena bonareinsis.

This bright autumn female chose my zinnias to fuel her journey south.

I tagged this monarch born in my yard in August. Scientists at the University of Kansas supply tags to interested "citizen scientists" to mark last-generation monarchs. Scientists and volunteers in Mexico watch for tagged butterflies arriving to the sanctuaries and relay tag numbers back to the staff at the university. Scientists learn more about monarch migration routes year after year through this tagging program.

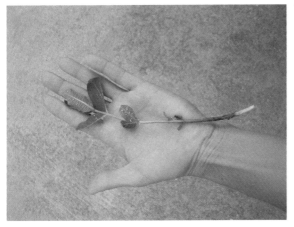

Throughout the summer, the female monarch usually deposits one egg on a milkweed plant. However, in late August a couple of years ago, a female laid thirty-one eggs on this monarch seedling below. This is called a "monarch dump." She may have laid so many eggs here since this seedling is tender and fresh whereas most milkweed in late summer is starting to dry up and die.

Not all monarchs migrate to Mexico. Monarchs residing west of the Rocky Mountains roost on trees along the California coast during the winter months. Only the monarchs from Canada and those living east of the Rocky Mountains in the U.S. migrate.

Monarchs from as far away as Ontario, Canada gather together and fatten up before flying more than 3,000 miles south. The southbound monarchs can hitch a ride on warm air currents called thermals, so they do not have to continually flap their wings during this arduous journey. Sometimes, hundreds of monarchs are seen flying together in the fall.

Up until 1975 no one was sure where the monarchs went. But flying overhead in an airplane that fall, scientists spotted orange areas in the Mexican mountains, which turned out to be the overwintering sites of hundreds of millions of butterflies. These twelve major roosting sites called sanctuaries are located about ninety miles west of Mexico City in the Mexican state of Michoacán.

Millions upon millions of monarchs are able to rest in the cool and moist microclimate found in the oyamel forest from November until

March. The monarchs are so thick on the trees that limbs have broken under their weight. At times, roosting monarchs do fly off the trees to search for water in puddles and melting snow.

Sometime in late February or March, the monarchs in Mexico become active due to warming temperatures and longer days and their reproductive system fully develops. Flying starts; mating begins. The females deposit their eggs on milkweed in Texas. They die soon after and their offspring are the monarchs that I see in my garden in May. Monarchs repopulate the U.S. and Canada as they journey north and lay eggs.

Research is still being done to figure out how the last-generation monarchs in Canada and in the U.S. find their way to such remote overwintering grounds in the Mexican mountains. Some scientists believe that as days shorten and weather cools up north, monarchs are drawn to travel south and follow a route guided by the angle of the sun. And maybe the sun guides them back up north in the spring.

While monarch populations can fluctuate throughout the years, many scientists are very concerned, even worried, about the future of the butterflies. One major issue is shrinking overwintering sites because of logging in Mexico. Other scientists are concerned about the effects of climate change and infestations of deadly insects found in the fir forests. With smaller and smaller sanctuaries, they say that the phenomenon of monarch migration is at risk.

Worries also surface on the U.S. side as fields of milkweed are destroyed due to urban development and use of agricultural herbicides. Scientists advocate for preservation of the sanctuaries in Mexico and encourage gardeners across the U.S. to plant milkweed in their communities to help restore significant amounts of monarch habitat loss.

As families and individuals across the country become educated about monarch life and habitat loss, they are becoming advocates for this beloved butterfly. More and more people are growing milkweed in their own backyards and are refraining from using herbicides and pesticides. Ordinary citizens are becoming the saving grace for the monarch. I know I will continue to kneel down each spring, run my fingers across the earth, and celebrate the appearance of the milkweed, which calls forth the magnificent monarchs.

How to raise a monarch
To raise your own monarch, you will need:
- Milkweed
- Monarch egg or caterpillar
- Large plastic container with screen
- Small plastic container to hold water

Check your local nursery for milkweed native to your area. Plant some milkweed and keep your eye out for the monarchs. Even if you do not see a monarch, check your plants every day or so for eggs. Once you find an egg, cut the leaf and place it in a tiny bit of water (or set the leaf on a damp paper towel in a small plastic cup with the lid closed). After the caterpillar hatches, place the leaf and the tiny caterpillar onto a milkweed cutting kept fresh in water in a large container. The caterpillar will crawl onto the fresh milkweed by itself and will begin eating the plant.

If your caterpillar needs more food, add some more leaves. Once a caterpillar eats for ten to fourteen days, it is nearing full size. Clean out the container and make sure that the bottom stays dry so mold and mildew do not form.

When full-grown, the caterpillar may climb to the top of your container to make its chrysalis. Keep your container still while the caterpillar is in the "J" position and when forming its wonderful chrysalis.

You can open the container and remove all contents a day after the chrysalis forms. After about ten days you may notice the chrysalis start to change color from jade to black. You probably will see the wing's design through the chrysalis.

The butterfly will emerge in a day or two, most likely in the early morning. Once a butterfly is born and has been drying for several hours, it is ready to be released. You can bring the container outside and remove the lid. It may take its first flight all by itself. If you want to assist, you can place your index finger between its legs so it crawls onto your index finger.

If it is ready, it will take off right away. If it is not, you can set the butterfly on a tree branch or bush. Sometimes it can take up to a day to fly depending on how cool it is outside. Delay the release if it's raining or dark outside. Enjoy!

Midwest timeline:
April - May: Milkweed surfaces in the garden.
May - June: First monarchs arrive.
September: Last generation of monarch butterflies born.
Monarch butterflies fatten up and migrate to Mexico.

Fun Fact: Monarch butterflies (*Danaus plexippus*) have been named the state insect of Illinois as well as several other states.

Websites:
Journey North: www.learner.org/jnorth. This website includes monarch sightings and maps and provides weekly updates about monarch activity. I check this website in late winter to see where the monarchs are and when I can expect them to arrive in my area. I post the date of the first monarch and eggs I see in my backyard as well as the date of their departure to Mexico.
Monarch Watch: www.monarchwatch.org. This University of Kansas website is dedicated to monarch education, conservation, and research.

Books:
Chasing Monarchs: Migrating with the Butterflies of Passage by Robert Michael Pyle. Mariner Books, 2001.
Four Wings and a Prayer: Caught in the Mystery of the Monarch Butterfly by Sue Halpern. Pantheon, 2001.

Comments that I received via e-mail from people who attended one of my backyard presentations:

- Thanks again for giving us the opportunity to learn about and witness the life cycle of the butterfly. This has been an amazing experience. I just wanted to share with you the pictures we took of Eduardo, my butterfly. —John
- Just wanted to say thank you again for the informational evening in your yard on butterflies. So many things one takes for granted in this world and you brought to my attention amazing things about butterflies—especially the monarch. I am definitely going to try and create a habitat in my yard. My grandson is taking good care of his caterpillar. —Mary Ellen
- I'd just like to thank you again for taking the time and effort to teach us about butterflies on my birthday weekend. What they represent in my life is incredibly meaningful, and the surprise of getting to witness and learn about them was something that has immediately gone down in my list of favorite memories of my entire life. —Marie
- Julianna came home from school today and said the science teacher thought it would be great if Julianna brought in a caterpillar for the class to watch its transformation. Again, thanks for engaging Julianna in this process! —Christine
- My caterpillar is big and beautiful, but I guess unless he decides to start the change tonight I will need to grab more milkweed from you. Thanks for this fun experience. I have a friend in her 80s who would really enjoy this. —Denise
- Henry (Henrietta) the monarch butterfly has left us! Thank you for that experience. I've been talking about it all weekend. —Chris

CPSIA information can be obtained at www.ICGtesting.com
Printed in the USA
LVIW01n2332260717
542800LV00002B/13

Science / Nature / Butterflies

RAISING MONARCHS
Caring for One of God's Graceful Creatures

by SUE FOX McGOVERN

Sue Fox McGovern and one of her monarchs

Have you ever wondered how to raise monarch butterflies? Author and butterfly farmer Sue Fox McGovern takes you through all of the steps from egg to cate~~rpillar~~ chrysalis to ~~butterfly~~ ~~can~~ be done in

SUNBURY
PRESS
USA $16.95

Published by
Sunbury Press, Inc.
Mechanicsburg, PA
MADE IN THE USA

www.sunburypress.com

ISBN 978-1-62006-246-
5169

9 781620 062463

Fox Has A Box

by H.P. Gentileschi

Reading Level A

X